Lab Manual – Universal Fluid Power Trainer
HSV1/MSOE01

Introduction to Hydraulics for Industry Professionals

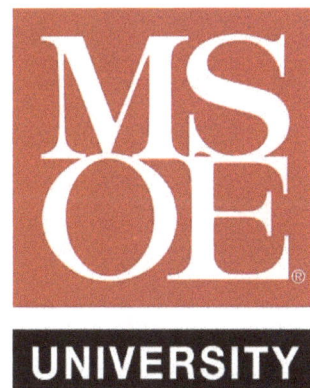

MS OE
UNIVERSITY

www.msoe.edu/seminars

Lab Manual
Universal Fluid Power Trainer (UFPT)
Introduction to Hydraulics for Industry Professionals

ISBN: 978-0-9977816-0-1

Copyright © 2016 by
Milwaukee School of Engineering

Printed in the United States of America

Lab Manual – Universal Fluid Power Trainer
Introduction to Hydraulics for Industry Professionals

Lab Manual-UFPT

How to Use Me Safely?

1- Emergency Stop:
- Locate the emergency stop.
- Press the emergency stop immediately if you feel any dangerous situation e.g. leakage, bad smell, system instability, unusual noise

2- Warning Light: consult the instructor if any of red led lights is turned ON.

3- Consult the Instructor: if you are not sure about any thing.

Please, no use of flash drives

Safety Regulations, Contd.

4. Safety Glasses: Wear the safety glass and shoes where applicable.

5. Hydraulic Hoses: Use **proper length &** respect minimum **bend radius.**

6. Quick Disconnect: Make sure it is perfectly connected **"hear the click".**

7. Trapped pressure: Connect high pressure measurement hose to the tank first, then to the point at which the pressure is trapped.

8. Accumulator (stored energy hazard): Isolate the accumulator whenever it is not needed.

9. Pressure Measurement Gauges: P4 assigned for tank pressure only using female-female hose between the tank header and P4.

10. Drain: make sure drain lines are plumbed where needed.

11. Flow-meter: is a unidirectional flow meter, follow the arrow.

12. Turning ON any power : Warn all team members first.

13. Circuit (Hydraulic/pneumatic/24V) Modification: turn the corresponding power button Off first.

Safety Regulations, Contd. For courses other than Introduction

---

14. Cables & Sockets: Keep the unused electrical sockets covered.

15. Cables & Sockets: Do not stretch or pull from wires & make sure it is correctly fit.

16. Servo Valve Cable: Servo valve cable contains DC/DC convertor.

17. Banana Jacks Cables: Use proper length of cables.

18. Electrical 24V Circuits: In order not to get confused and to avoid a short circuit, the good advice is to connect one vertical line at a time.

19. Pressure Sensors: make sure it is tightened enough to sense the pressure. High pressure sensors are numbered (1/4 – 3/4) and one low pressure is tagged (1/4).

20. Pneumatic Power: turn air valves off before disassemble air hose.

21. Compressor(stored energy hazard): Before turning the compressor ON, make sure air valves are off.

22. Follow up the "lab procedure".

23. Follow the machine "Machine Startup Procedure".

24. Follow the machine "Machine Shutdown Procedure".

Lab Procedure

1. Form work groups. A group works together till the end of the class..
2. Listen to the instructor orientation. Ask questions if needed.
3. Read the lab instructions line by line even after the orientation.
4. Read schematics (Electrical/Hydraulics) and work instructions.
5. Build the system as per the step-by-step given instructions.
6. Operate the circuits and follow the given instructions.
7. Recorded your observations as per the exercise instructions.
8. Analyze the data and answer the posted questions.
9. Disconnect the circuit and put down hoses and measuring instruments at the end of every lab session.
10. Store back the components as per its drawer identification.

Lab Procedure, Contd.

Hydraulic Power Supply

Adjustment of PRV at the beginning of every exercise

Matlab RT Models
1-Activating, 2-Connecting 3-Running

Matlab

Current Folder: C:\11-Common Matlab\Models Real Time

Simulink ▶ 100 External

Machine Startup Procedure

1. Casters: Lock the casters to prevent accidental move of the unit.
2. Uncoil the power cords and plug the unit to the electrical wall outlet.
3. Take the hoses out of the hose bin and hang them on the side hooks.
4. Turn ON the electrical control panel.
5. Turn ON the HMI (only when needed).
6. Turn ON the printer (only when needed).

6

Machine Shutdown Procedure

1. Make sure cylinders are fully retracted.
2. Discharge the accumulator from the manual discharge valve.
3. Discharge the compressor by opening one of the air valves.
4. Turn OFF the pump, air power, and control power.
5. Shutdown the computer from the windows.
6. Turn OFF the mouse and store it back in drawer "A"
7. Turn OFF the 24 Volt power supply.
8. Turn OFF the printer.
9. Set toggle switches to 'Pot" position.
10. Set the potentiometers to "min".
11. Unplug the power cord and hang it under the counter.
12. Unplug any remaining electrical connections.

7

Machine Shutdown Procedure, Contd.

13. Disconnect hydraulic hoses, cover their ends by dust caps, and bring the store hoses back to the hose bin.
14. Cover all hydraulic components with dust caps.
15. Store the components as per its drawer identification.
16. Please, clean the safety glasses and store them.
17. Please, clean spilled oil.

8

MSOE – Fluid Power and Motion Control Professional Education

Lab01- Energy Losses in a Hydraulic System

Objectives:

- Practice the **first use of the trainer** and date collection from measuring instruments.
- **Observe the pressure drop** in a hydraulic circuit.
- **Calculate the power losses** in a hydraulic circuit.
- **Calculate the Reynolds number.**
- Define the **flow type** in a hydraulic line.

Step 1: Prepare Components:

MSOE#-X	Component Name	QTY	Drawer #
N27	Pressure measurement hose	4	B
	Manually Operated Valve Bank	1	D6
N16-TX-2/4	Flow meter (Medium Scale)	1	D2

Exit

0

Step 2: Adjust the PRV to 500 psi:

1. Connect the pressure measurement hose as shown to P1.
2. Set the Pp and Qp toggle switches to "Pot" positions.
3. Adjust pump potentiometers, Pp and Qp, to maximum positions.
4. Open the PRV halfway.
5. Turn ON the hydraulic power.
6. Close the PRV gradually until the pressure gauge reads the set value.
7. Turn OFF the hydraulic power.

Exit

1

Step 3: Hydraulic Circuit:

1. Build the shown hydraulic circuit with long hoses to get better sense of the losses.
2. Turn the hydraulic power ON.
3. Shift the manual directional valve to either direction.

Step 4: Procedure:

- Measure and record the data in the following table.

Flow rate Q (gpm)	Pressure Location	Pressure (psi)
-	P1	
-	P2	
-	P3	
	P4	

Calculations and Analysis:

- Calculate the total pressure drop in the system = P1 – P4 =
- Calculate the total power losses due to the total pressure drop in the system:

$$\text{Power (HP)} = \frac{\Delta p(\text{psi}) \cdot Q(\text{gpm})}{1714} =$$

- Calculate the Reynolds number for the flow in the connecting hoses.

$$R_e = \frac{3164.2 \times Q[gpm]}{\nu[Cst] \times D[in]}$$

Given :

ν = Kinematic Viscosity = 32 cSt

D = Hose Diameter (3/8) in

- What is the type of flow? [Laminar – in Transition – Turbulent]

Step 5: Lab Summary:

- Working pressure is not just the pressure required to support the load.
- Wasted energy within the system may result in considerable increase in the working pressure at the pump.
- If the flow is not laminar, what is your suggestions to resolve this issue?

Groups	G1	G2	G3	G4
PSID$_{total}$ [psi]				
Wasted Power [HP]				
Flow Type				

5

Lab02- Power Distribution in a Hydraulic System

Objective:

This experiment is to practice analyzing pressure distribution in a hydraulic system. Another goal is to calculate the energy saved by using pump-controlled motor speed control instead of valve-controlled speed control.

Step 1: Prepare Components:

MSOE#-X	Component Name	QTY	Drawer #
N27	Pressure measurement hose	4	B
	Manually Operated Valve Bank	1	D6
N07	Throttle-Check Valve	1	D2
N16-TX-2/4 N16-TX-3/4	Flow meter (Medium Scale)	2	D2
N06	Direct-acting pressure relief valve	1	D4

Step 2: Adjust the PRV to **750** psi:

1. Connect the pressure measurement hose as shown to P1.
2. Set the Pp and Qp toggle switches to "Pot" positions.
3. Adjust pump potentiometers, Pp and Qp, to maximum positions.
4. Open the PRV halfway.
5. Turn ON the hydraulic power.
6. Close the PRV gradually until the pressure gauge reads the set value.
7. Turn OFF the hydraulic power.

Step 3: Case A: (Control using valves)

1. Build the shown hydraulic circuit.
2. Keep the throttle valve fully closed
3. Turn the hydraulic power ON.
4. Shift the DCV so that the motor will not spin because throttle valve is fully closed.
5. Use the external PRV to adjust the system pressure to be 300 psi
6. Open the throttle valve to adjust the motor speed at 500 rpm.
7. Record the following measurements in the table shown below.
8. Turn the hydraulic power supply Off.
9. Calculate the power distribution as follows:

	P_1 psi	P_2 psi	P_3 psi	P_4 psi	Q_{RV} gpm	Q_m gpm

$$1 - \text{Power Throttle (kW)} = \frac{(P_3 - P_4)(\text{psi}) \cdot (Q_m)(\text{gpm})}{1714 \times 1.36} =$$

$$2 - \text{Power Motor (kW)} = \frac{(P_2 - P_3)(\text{psi}) \cdot (Q_m)(\text{gpm})}{1714 \times 1.36} =$$

$$3 - \text{Power PRV (kW)} = \frac{(P_1 - P_4)(\text{psi}) \cdot (Q_{RV})(\text{gpm})}{1714 \times 1.36} =$$

$$4 - \text{Power Pump (kW)} = \frac{(P_1)(\text{psi}) \cdot (Q_m + Q_{RV})(\text{gpm})}{1714 \times 1.36} =$$

Is the pump power = Throttle + Motor + PRV?

If not, where is the difference?

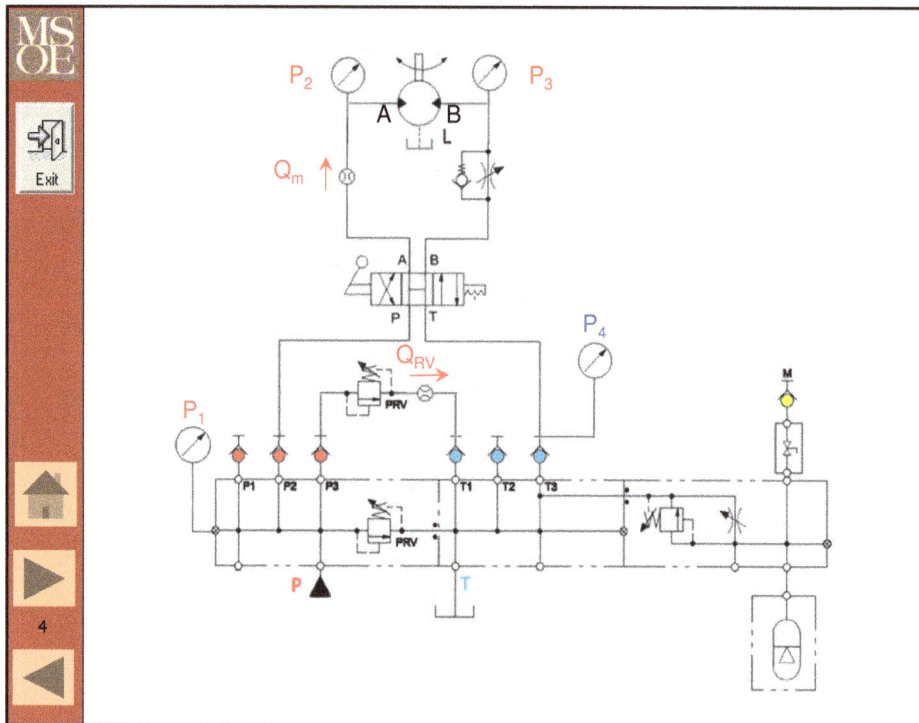

Step 4: Case B: (Valve-less Control)

1. Rebuild the hydraulic circuit as shown.
2. Turn the hydraulic power ON.
3. Adjust the pump displacement so that the motor runs at 500 RPM
4. Record the following measurements: P1 (psi) =   Qp (gpm) =
5. Turn the hydraulic power supply Off.
6. Calculate the pump power.

$$\text{Power Pump (kW)} = \frac{(P_1)(psi) \cdot (Q_P)(gpm)}{1714 \times 1.36} =$$

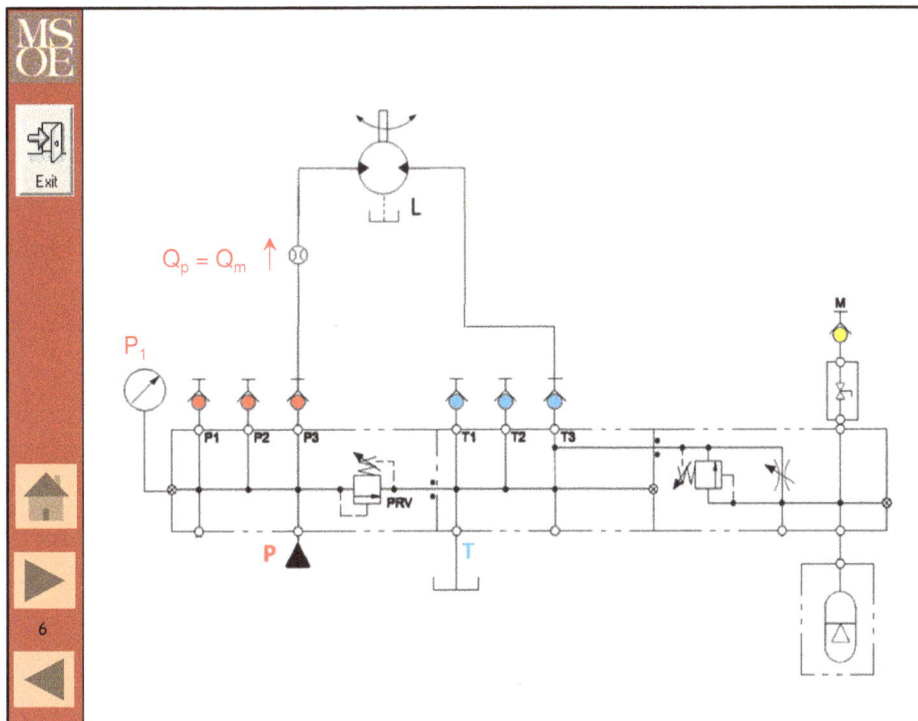

Step 5: Energy Saving

Calculate the energy saving by using valve-less control approach as follows:

Energy Case A(kJ) = Power Pump (kW)x60x60 =

Energy Case B (kJ) = PowerPump (kW)x60x60 =

Energy Saving (kJ) = EnergyA(kJ) − EnergyB (kJ) =

Step 6: Lab Summary

Groups	G1	G2	G3	G4
Case A: Pump Power (Valve Control)				
Case B: Pump Power (Valve-less Control)				
Energy Saving (kJ)				

Lab03- Valve Coefficient Development

Objectives:
- Measure pressure drop across directional control valves.
- Calculate the valve's coefficient at specific operating condition.
- Compare open center with closed center type of valve.

Step 1: Prepare Components:

MSOE#-X	Component Name	QTY	Drawer #
N27	Pressure measurement hose	3	B
	Manually Operated Valve Bank	1	D6
N16-TX-2/4	Flow meter (Medium Scale)	1	D2

Step 2: Adjust the PRV to 500 psi:

1. Connect the pressure measurement hose as shown to P1.
2. Set the Pp and Qp toggle switches to "Pot" positions.
3. Adjust pump potentiometers, Pp and Qp, to maximum positions.
4. Open the PRV halfway.
5. Turn ON the hydraulic power.
6. Close the PRV gradually until the pressure gauge reads the set value.
7. Turn OFF the hydraulic power.

Step 3: Hydraulic Circuit: Case A - (use open center valve)

- Build the shown hydraulic circuit and turn the hydraulic power ON.
- Shift the directional valve so that the flow passes from port A to port B of the valve.
- Record the flow and the pressure drop (P2-P3) across the valve.
- Turn the hydraulic power OFF.

- Case B - (use closed center valve)
- Build the shown hydraulic circuit and turn the hydraulic power ON.
- Invite the instructor before you continue. The instructor will ask a question.
- Shift the directional valve so that the flow passes from port A to port B of the valve.
- Record the flow and the pressure (P2-P3) drop across the valve.
- Turn the hydraulic power OFF.

Step 4: Lab Summary – calculate the valve coefficient

Specific Gravity = 0.9.

$$Q = C_D A \sqrt{\frac{2(p_2 - p_3)}{\rho}} = C_v \sqrt{\frac{\Delta P}{Sg}} \Rightarrow C_v = Q \sqrt{\frac{Sg}{\Delta P}}$$

Valve	G1	G2	G3	G4
C_{VPA} Case A: Open Center				
C_{VPA} Case B: Closed Center				

Lab04- Motion Control of Hydraulic Cylinder

Objectives:

- Practice how to control the motion of a hydraulic cylinder using open center valve and closed center valve and observe the difference.
- Observe the effect of the area ratio on both extension and retraction speeds.
- Calculate the area ratio of the cylinder.

Step 1: Prepare Components:

MSOE#-X	Component Name	QTY	Drawer #
N27	Pressure measurement hose	1	B
	Manually Operated Valve Bank	1	D6
N16-TX-2/4	Flow meter (Medium Scale)	1	D2
N16-TX-4/4	Flow meter (High Scale)	1	D2

Step 2: Adjust the PRV to 500 psi:

1. Connect the pressure measurement hose as shown to P1.
2. Set the Pp and Qp toggle switches to "Pot" positions.
3. Adjust pump potentiometers, Pp and Qp, to maximum positions.
4. Open the PRV halfway.
5. Turn ON the hydraulic power.
6. Close the PRV gradually until the pressure gauge reads the set value.
7. Turn OFF the hydraulic power.

Step 3: Hydraulic Circuit:

- Case A - (use open center valve)
- Build the shown hydraulic cylinder and turn the hydraulic power ON.
- Shift the directional valve to extend and retract the cylinder and observe the cylinder speed in both strokes.
- Answer question #1 in the lab summary.
- Return the valve to its neutral position. Observe the cylinder motion.
- Answer questions #2&3 in the lab summary and record the pump pressure.
- Shift the valve to extend the cylinder. At the middle of the cylinder extension stroke, record flow supplied to the piston side and the flow generated from the rod side.
- Answer question #4 in the lab summary.
- Shift the valve to retract the cylinder. At the middle of the cylinder retraction stroke, record flow supplied to the rod side and the flow generated from the piston side.
- Answer question #5 in the lab summary.
- Turn the hydraulic power Off.

- **Notes:**
- In the shown flow-meters position, direction of flow is kept the same, but readings are flipped based on the stroke.
- Measurements must be taken during cylinder movement at the middle of its stroke.

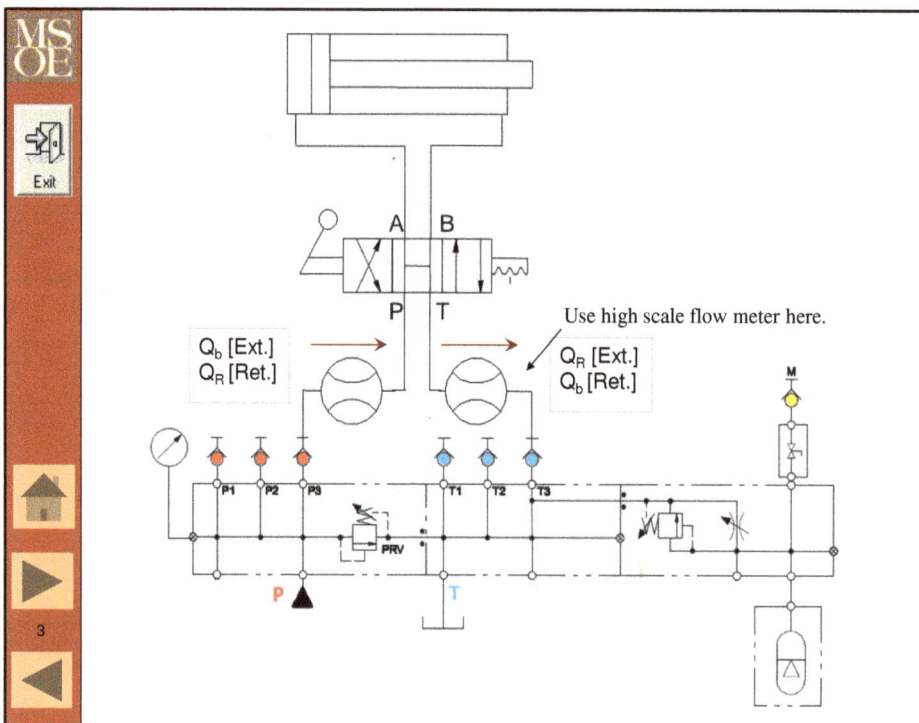

Use high scale flow meter here.

Q_b [Ext.]
Q_R [Ret.]

Q_R [Ext.]
Q_b [Ret.]

MSOE – Fluid Power and Motion Control Professional Education

20

- Case B - (use closed center valve)
- Repeat the steps but with the closed-center valve.

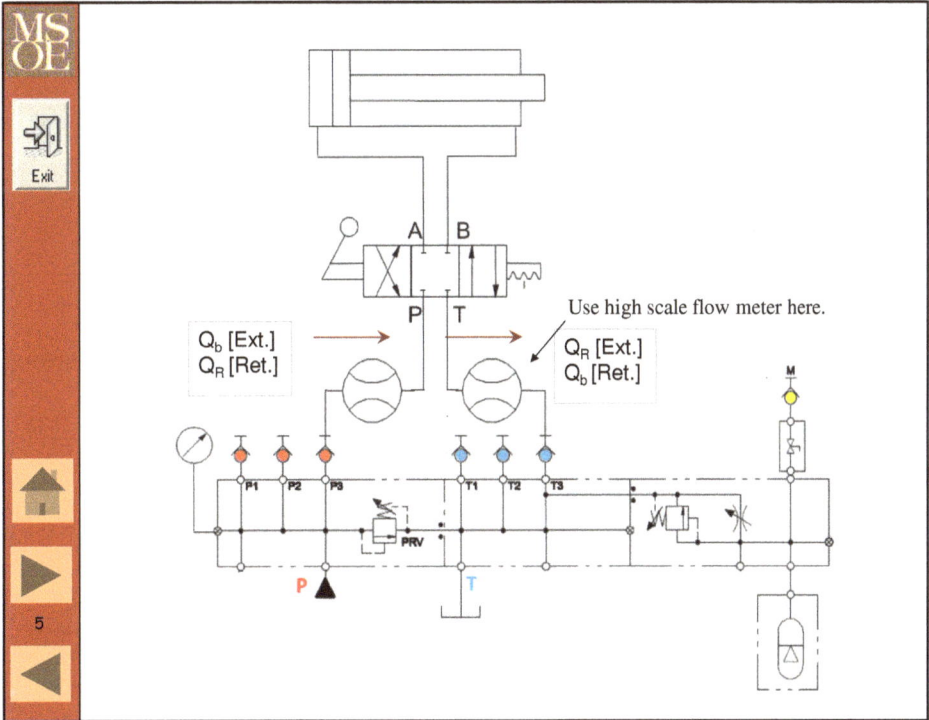

Exit

4

Exit

Use high scale flow meter here.

Q_b [Ext.]
Q_R [Ret.]

Q_R [Ext.]
Q_b [Ret.]

A B

P T

P1 P2 P3

T1 T2 T3

PRV

M

P

T

5

Step 4: Lab Summary

Valve	Case A - Open Center Valve			Case B - Closed Center Valve		
Readings	Valve P [psi]	Q_{blind} [gpm]	Q_{rod} [gpm]	Valve P [psi]	Q_{blind} [gpm]	Q_{Rod} [gpm]
Extension						
Neutral						
Retraction						

- Calculate the average cylinder area ratio
- [AR = A_{blind}/A_e = Q_{blind} / Q_{rod}] =

Group	G1	G2	G3	G4
AR				

Use your readings and observations to confirm the following lab summary:

1. Is extension speed < retraction speed?
2. Cylinder extended when the valve is in its neutral position?
3. Pump pressure reading at valve neutral position < back panel PRV setting?
4. During cylinder extension, ($Q_{blind} = Q_{pump}$)> Q_{rod}?
5. During cylinder retraction, Q_{blind} > ($Q_{rod} = Q_{pump}$)?

Question	Case A - Open Center Valve		Case B - Closed Center Valve	
1	☐ True	☐False	☐ True	☐False
2	☐ True	☐False	☐True	☐ False
3	☐ True	☐False	☐True	☐ False
4	☐ True	☐False	☐ True	☐False
5	☐ True	☐False	☐ True	☐False

Lab05- Control of Overrunning (Vertical) loads

Objective:

- Practice control of overrunning (vertical) load using different approaches, meter-out speed control, counterbalancing and use of pilot operated check valves.

Step 1: Prepare Components:

MSOE#-X	Component Name	QTY	Drawer #
N27	Pressure measurement hose	3	B
N07	Throttle-Check Valve	1	2
N08	Counter Balance Valve	1	4
N11	Pilot-Operated Check Valve	1	4
N19	T-Connection	1	2
	Manually Operated Valve Bank	1	D6

Exit

0

Step 2: Adjust the PRV to 500 psi:

1. Connect the pressure measurement hose as shown to P1.
2. Set the Pp and Qp toggle switches to "Pot" positions.
3. Adjust pump potentiometers, Pp and Qp, to maximum positions.
4. Open the PRV halfway.
5. Turn ON the hydraulic power.
6. Close the PRV gradually until the pressure gauge reads the set value.
7. Turn OFF the hydraulic power.

Exit

1

Step 3: Hydraulic Circuit:

Case A: Circuit to build: (**Meter-Out Control**)

- Build the shown hydraulic cylinder and turn the hydraulic power ON.
- Practice extending and retracting the cylinder.
- Observe the speed control resolution.
- Adjust the lowering speed to a relatively low speed.
- Measure P_2 and P_3 **during lowering** the load (in the middle of the stroke).
- Turn the hydraulic power OFF.
- Calculate the seal friction force as follows:

$$P_2A_2 + F = P_3A_3 + W \Rightarrow$$
$$F(lb) = P_3(psi) \times A_3(in^2) + W(lb) - P_2(psi) \times A_2(in^2) \Rightarrow$$
$$F(lb) = P_3(psi) \times 0.98125(in^2) + 35(lb) - P_2(psi) \times 1.76625(in^2) =$$

2

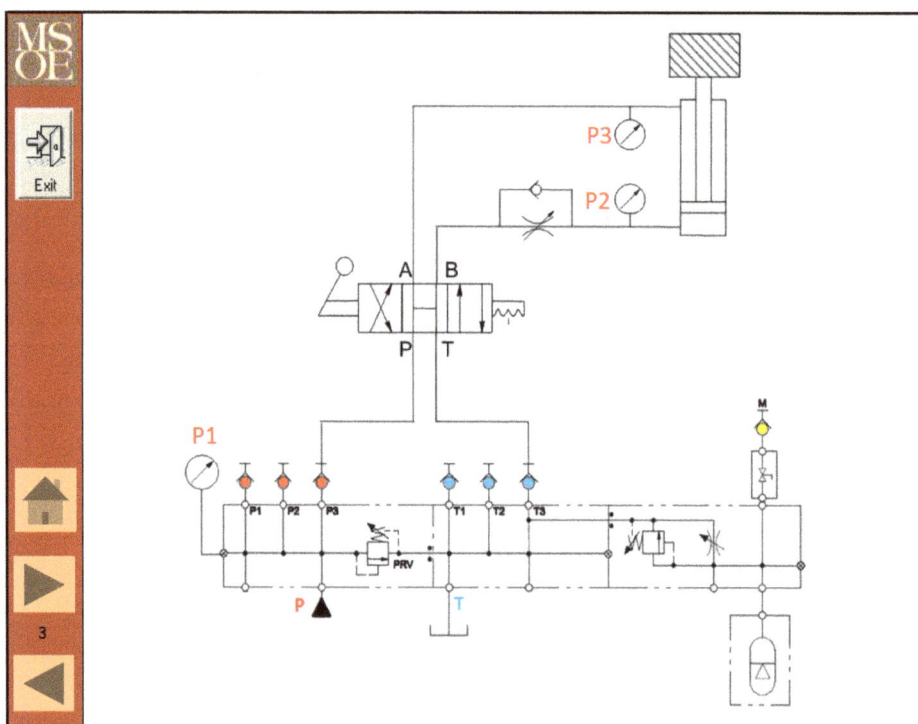

3

Case B: Circuit to build: (**Counterbalancing**)

- Build the shown hydraulic cylinder and turn the hydraulic power ON.
- Practice extending and retracting the cylinder.
- Observe the speed control resolution as compared to the use of meter-out.
- Turn the hydraulic power OFF.

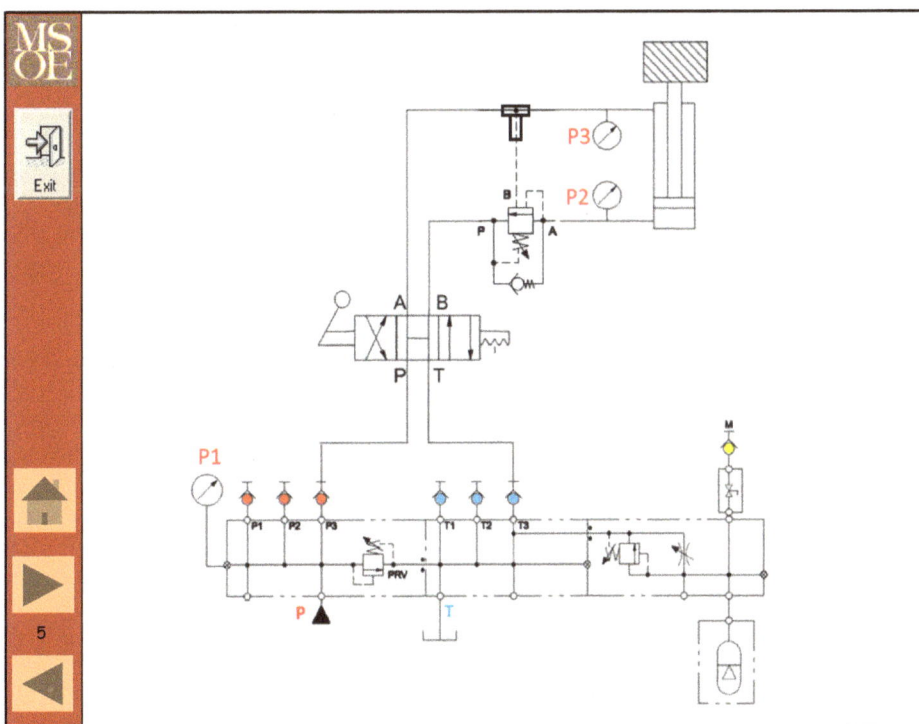

Case C: Circuit to build: (Pilot Operated Check Valve)

- Build the shown hydraulic cylinder and turn the hydraulic power ON.
- Practice extending and retracting the cylinder.
- Do you see any way to control cylinder speed?
- What is then the benefit of using this technique?
- Turn the hydraulic power OFF.

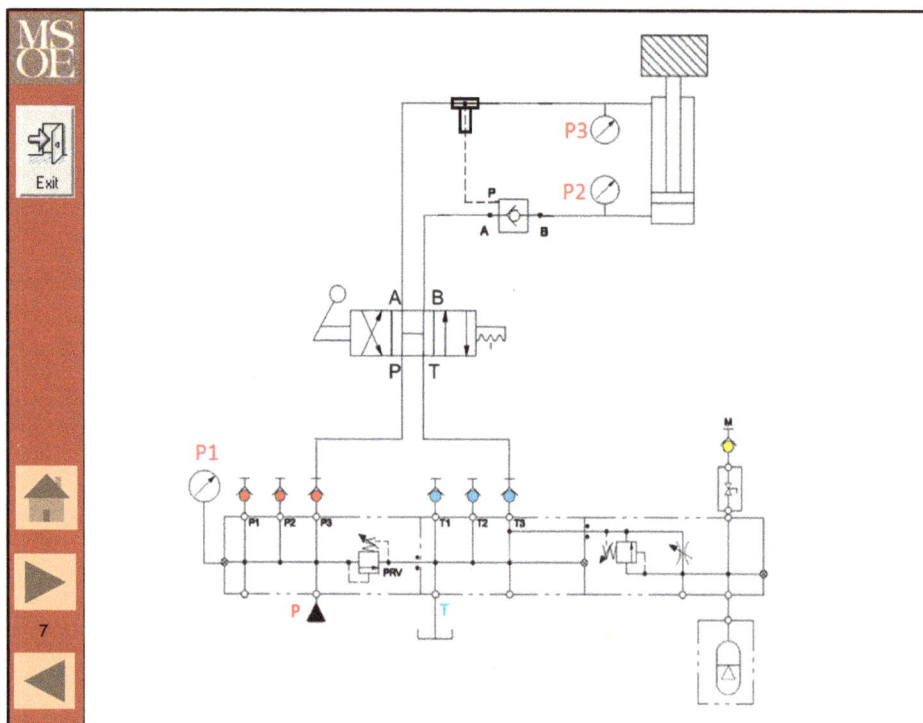

Step 4: Lab Summary

	Case A Meter-Out	Case B Counter Balancing	Case C Pilot Operated Check Valve
Method offer finest speed control resolution			
Method offer supporting overrunning load with leak free conditions			

Viscous Friction Force

Group	G1	G2	G3	G4
F(lb)				

Lab06- Speed Control of a Hydraulic Actuator

Objectives:

- Practice controlling the speed of a hydraulic cylinder using different approaches.

- Observe the differences in the working conditions using different approaches.

Step 1: Prepare Components:

MSOE#-X	Component Name	QTY	Drawer #
N27	Pressure measurement hose	3	B
N07	Throttle-Check Valve	1	2
N16-TX-1/4	Flow meter (Low Scale)	1	D2
	T-Connection	1	2
	Manually Operated Valve Bank	1	D6

Step 2: Adjust the PRV to 500 psi:

1. Connect the pressure measurement hose as shown to P1.
2. Set the Pp and Qp toggle switches to "Pot" positions.
3. Adjust pump potentiometers, Pp and Qp, to maximum positions.
4. Open the PRV halfway.
5. Turn ON the hydraulic power.
6. Close the PRV gradually until the pressure gauge reads the set value.
7. Turn OFF the hydraulic power.

Step 3: Hydraulic Circuit:
- Build the circuits shown in the following 3 cases.
- Turn the power off when you get to change the circuit.
- In each case, adjust the throttle valve to set the motor rpm as shown in the table.
- After setting the rpm, record the pressure readings.
- Share your observation with the instructor.

Case A: (**bleed off speed control**)

Motor RPM	Pump = P1 psi	Motor = P2 psi	Motor =P3 psi
400			
600			

- Does this method offer the least working pressure in the system? (True – False)
- Is this method recommended for constant loading conditions? (True - False)

Case B: (Meter-In control)

P2   P3   A   B   L   P1

Motor RPM	Pump = P1 psi	Motor = P2 psi	Motor =P3 psi
400			
600			

- Does this method offer relatively stabilized speed control? (True – False)
- Can this method hold an overrunning load? (True – False)

A B P T — M — PRV — P — T

Case C: (Meter-Out control)

P2   P3   A   B   L   P1

Motor RPM	Pump = P1 psi	Motor = P2 psi	Motor =P3 psi
400			
600			

- Does this method offer the highest working pressure in the system? (True – False)
- Can this method hold an overrunning load? (True – False)

A B P T — M — PRV — P — T

UFPT-Lab06

MSOE – Fluid Power and Motion Control Professional Education

30

Lab07- Boosting Speed of a Hydraulic Cylinder

Objectives:

- Practice boosting the speed of a hydraulic cylinder using different techniques.
- Observe and record the cylinder extension speed in each case.
- Calculate the stored oil volume in the accumulator and its gas pre-charge pressure.

Step 1: Prepare Components:

MSOE#-X	Component Name	QTY	Drawer #
N27	Pressure measurement hose	1	B
	Manually Operated Valve Bank	1	D6
N12	Check Valve	1	D2

Step 2: Adjust the PRV to 650 psi:

1. Connect the pressure measurement hose as shown to P1.
2. Set the Pp and Qp toggle switches to "Pot" positions.
3. Adjust pump potentiometers, Pp and Qp, to maximum positions.
4. Open the PRV halfway.
5. Turn ON the hydraulic power.
6. Close the PRV gradually until the pressure gauge reads the set value.
7. Turn OFF the hydraulic power.

Step 3: Hydraulic Circuit: Case A: (Conventional Circuit)

- Make sure the accumulator is isolated from the system (hand face right).
- Build the shown circuit and turn the hydraulic power supply ON.
- Shift the directional valve to fully retract the cylinder.
- Extend the cylinder and measure the time for full extension stroke: T(s) =
- Turn OFF the hydraulic power.
- Calculate and record the cylinder speed: Vext [in/s] = 10 in / T (s) =

Step 4: Hydraulic Circuit: Case B: (Regenerative Circuit)

- Make sure the accumulator is isolated from the system (hand face right).
- Build the shown circuit and turn the hydraulic power supply ON.
- Shift the directional valve to fully retract the cylinder.
- Extend the cylinder and measure the time for full extension stroke: T(s) =
- Turn OFF the hydraulic power.
- Calculate and record the cylinder speed: Vext [in/s] = 10 in / T (s) =

Step 5: Hydraulic Circuit: Case C: (Use of Accumulator)

- Make sure the accumulator is connected to the system, hand (1) face up.
- Make sure the accumulator manual discharge valve (2) is fully closed.
- Make sure the accumulator PRV (3) is fully closed.
- Build the shown circuit and connect pressure gauge to the test point (4)
- Turn ON the hydraulic power supply.
- Shift the valve to fully retract the cylinder.
- Extend the cylinder and measure Time for full extension stroke: T(s) =
- Calculate and record the cylinder speed: Vext [in/s] = 10 in / T (s) =

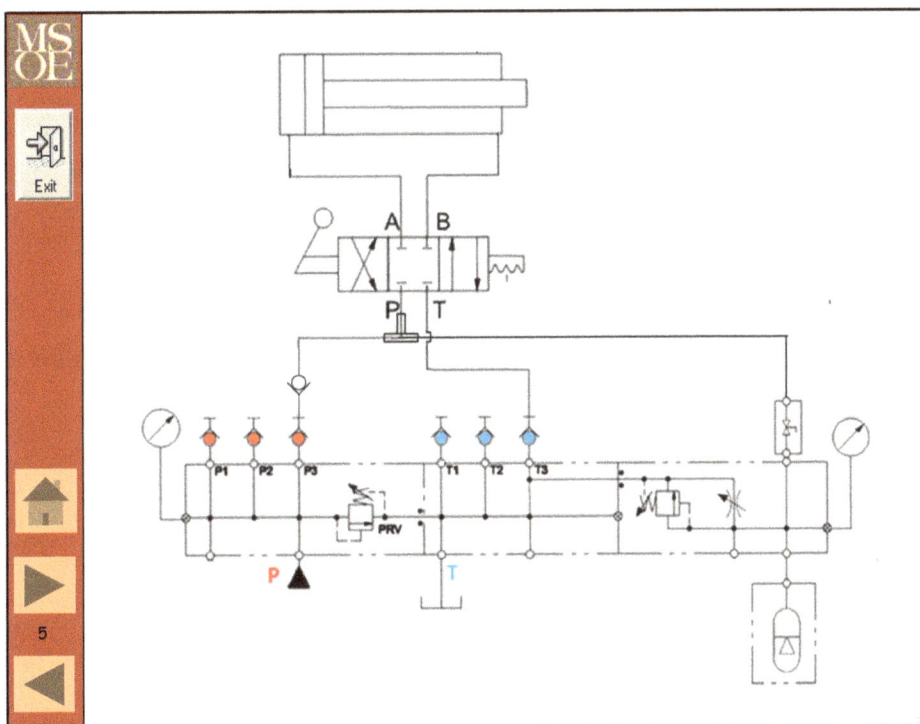

Step 6: Hydraulic Circuit: (Calculate the Accumulator pre-charge pressure)

- Fully retract the cylinder.
- Wait to give time to charge the accumulator and read 650 psi pump pressure.
- Turn OFF the hydraulic power.
- Make sure the accumulator pressure is not dropping OR consult the instructor.
- Extend/retract the cylinder until the stored energy in the accumulator is totally consumed and the cylinder stops.
- Measure the total extension/retraction travel =   in,
- Turn the hydraulic power ON.
- Fully retract the cylinder then turn power OFF.
- Isolate the accumulator by making hand (1) face right.

Calculations and Analysis:

1- Calculate the accumulator charge oil volume:

$\Delta V \text{ [in}^3] = Ap \quad \times \text{ Total Extension Travel} + Ae \quad \times \text{ Total Retraction Travel}$

$= 1.76625 \times \qquad + 0.98125 \times$

2- Calculate the pre-charge pressure based on isothermal analysis:

$P_1 V_1 = P_2 V_2 \Rightarrow P_1 = P_2 V_2 / V_1$

$P_1 = \dfrac{P_2[V_1 - \Delta V]}{V_1} = \dfrac{650[58 - \Delta V]}{58} = \dfrac{650[58 - \quad]}{58} = \qquad \text{psi}$

$P_{1=?}$, V_1 , $P_2 V_2$

Step 7: Lab Summary

Cylinder Speed Calculation

Groups	G1	G2	G3	G4
Extension Speed (in/s)				
Conventional				
Regenerative				
Accumulator				

Accumulator Calculation

Group	G1	G2	G3	G4
Delta V				
P_1				

Lab08- Sequence Control

Objective:

- Practice building sequence function of two hydraulic cylinders using sequence valve.

Step 1: Prepare Components:

MSOE#-X	Component Name	QTY	Drawer #
N27	Pressure measurement hose	1	B
N19	T-Connections	2	2
	Manually Operated Valve Bank	1	D6
N09	Sequence Valve	2	D4

0

Step 2: Adjust the PRV to 600 psi:

1. Connect the pressure measurement hose as shown to P1.
2. Set the Pp and Qp toggle switches to "Pot" positions.
3. Adjust pump potentiometers, Pp and Qp, to maximum positions.
4. Open the PRV halfway.
5. Turn ON the hydraulic power.
6. Close the PRV gradually until the pressure gauge reads the set value.
7. Turn OFF the hydraulic power.

1

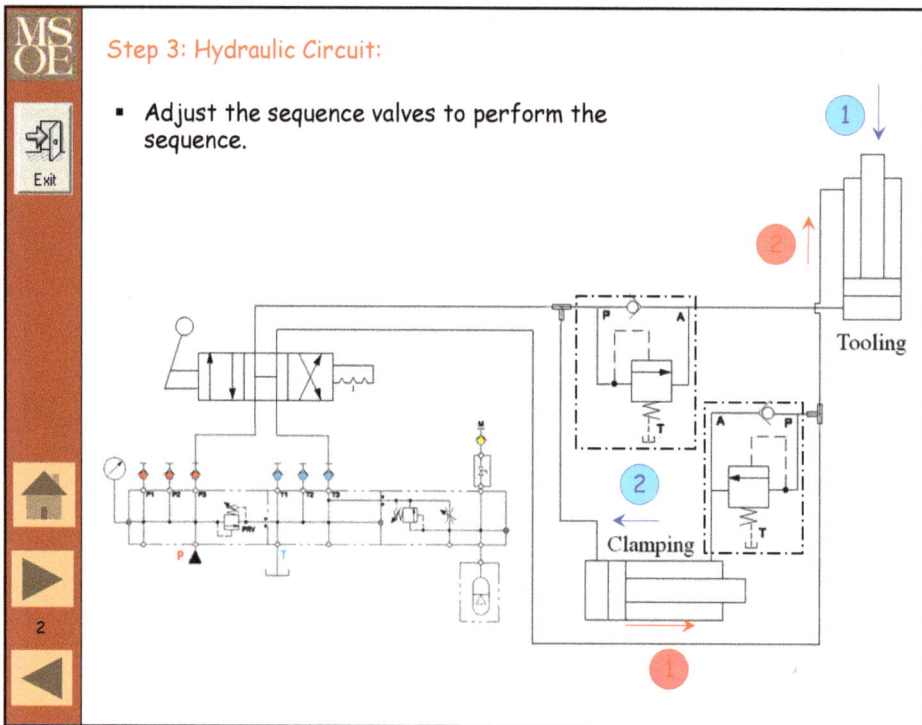

Step 3: Hydraulic Circuit:

- Adjust the sequence valves to perform the sequence.

www.ingramcontent.com/pod-product-compliance
Lightning Source LLC
Chambersburg PA
CBHW050241220326
41598CB00047B/7472